Memories of Kerry's Railways
A. H. Vaughan

Killarney had been a tourist attraction before the railway came, but in 1846 the promoters of the Killarney Junction Railway (KJR) saw the benefit to developing tourism and the prosperity of Killarney by having a railway from the main line at Mallow – when that was built. They also built a magnificent hotel to accompany the station. This is the station front in April 1979.

© A. H. Vaughan, 2022
First published in the United Kingdom, 2022,
by Stenlake Publishing Ltd.
www.stenlake.co.uk
ISBN 978-1-84033-938-3

The publishers regret that they cannot supply
copies of any pictures featured in this book.

Printed by
Claro Print, Office 26, 27, 1 Spiersbridge Way,
Thornliebank, Glasgow G46 8NG

Acknowledgements

I must thank these good men for their help with supplying vital details of Kerry railways: Hassard Stackpoole, Gerry McMahon, Patrick Ronan, Dermott Mullane, Ian Lake, Herbert Richards and Richard Norton.

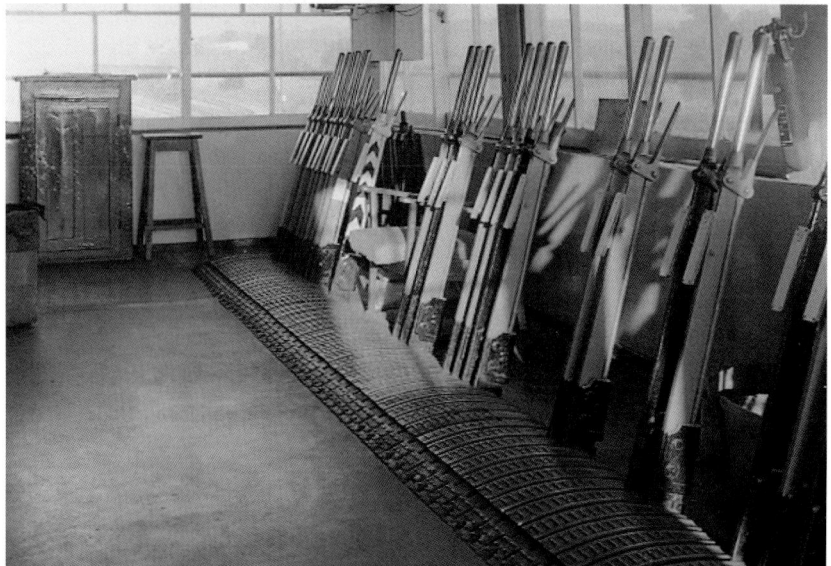

Right upper: *Gloucester Castle* with the Up 7.30 Carmarthen 'Red Dragon' passing Challow Signal Box. June 1960.

Right lower: Witham Signal Box lever 1975.

Introduction

Steam railways were my first love. It began in 1945 when I was four years old and was able to walk the mile to the playing field with my big sister. The Southern Railway was in a cutting beside the field and steam engines came thrashing up the hill from Reading. The engines were old and their exhaust seemed to 'miss a beat' as they came blasting up the grade.

I started my railway education on the Great Western, aged 5 in 1946. I grew up with steam engines and signal boxes because of the very kind and patient railwaymen who allowed me into their world. I learned the very basic routine of signalling in an 8-lever signal box, on the four track main line the London side of Reading. That was in 1951. I had often ridden on locomotive footplates of the Reading Station Pilot engine for 1951-53. I watched eagerly how they worked the engine and I listened to the tales that went between the driver and the fireman. Ten-platform Reading Station became my second home. I was allowed on the footplates of the engines that served as Station Pilot. By the time a driver offered me the chance to drive a little I knew exactly what to do – and did it well. The close attention I paid to the work in the engine's cab would have surprised any of my schoolmasters for whom I was a dreamy pain.

In 1953 my father moved the family into the deep countryside at the west end of Berkshire. Challow became my station, between Didcot and Swindon. I helped the porters with their work and rode on the daily shunting engine – whenever I could – and was on first name terms with one driver in particular. He offered me the chance to drive his little tank engine which I did at once, quite competently, because I had watched how the work was done on the Reading Pilot. One signalman in particular at Challow took me under his wing and, giving me the book of signalling Regulations, he bade me read it and coached me through it page by page. The Regulations were complicated, giving instructions for how to deal with every possible situation that might arise. A short time before my 15th birthday, the headmaster of the school told me that 'as you will be 15 before the school comes back after the Christmas holiday and as 15 is the minimum leaving age and there's nothing more we can do for you here – don't come back.' I remember his words well and was glad to leave.

Instead of at once asking for a lad porter's job at Challow – I joined the regular army and a trained to become 'an N.C.O and Warrant Officer of the future'. I wasn't bothered about all that but there would be regular journeys to and from Challow and Plymouth – free train rides in the stopping train to Reading and the 1.30 Paddington – Plymouth express from Reading behind magnificent 'King' class locos. After 4½ years uneventful years in the Army, 1956 – 60, I became an employee of British Railway Western Region, a lad porter, at Challow on 13th September 1960. In 1961 I became a signalman. I rode thousands miles on the footplates of goods and express trains. In January 1966 the last steam engine ran on Western Region. That was a great sorrow.

In 1970 I was made redundant twice but I was very lucky to meet a wonderful girl, Susan O'Sullivan. She had deep roots in County Kerry and changed my life. We were married in 1972. In 1973 I was redundant again and left Oxford for what was to be my last signal box – Witham. Also that year I bought a derelict cabin at Meanus, Killorglin. In September 1975 I resigned from Witham. Susan and I crossed the sea to Ireland and a new, free, life to sink or swim by what we could earn for ourselves. Our memoir of those years, just published, is called *Money Enough for the Winter*.

During the drive from Rosslare to Killorglin I was excited to see mechanical signalling. I could hardly wait to explore the railways of Ireland with my trusty Rolleiflex. We had a cabin to rebuild but I fitted in the photography – developing and printing without running water or electricity for a few years and used a neighbours garage for a dark room.

Killarney Station and bay platform and a well filled goods yard, 2nd October 1974.

Killarney Station from the embankment of the line to – or from – Tralee, photographed in October 1978. The extension to Tralee was built by the Tralee & Killarney Railway Company, incorporated by Act of Parliament on 15th August 1853 and opened on 18th June 1859. The business people of Tralee wanted a railway for the obvious reason of not having to ride a horse 20 miles to catch a train but also to have access to the electric telegraph which, since June 1853, had given Killarney businessmen an advantage in trading.

A train is signalled from Tralee, 4th April 1979. The train will go into the reversing siding and will reverse into the platform. The left-hand arm of the bracket signal as we look at it, when cleared, routes to the Mallow line. The other indicates that the points are set to go into the Reversing Siding in order to take the route for Tralee.

The railwaymen I met as I went around photographing were always very kind and understanding. The Killarney Signal Box was built by the Saxby & Farmer contractors and photographed in October, 1978. The signalman saw me photographing and like a good man invited me inside.

The Killarney signalman and his levers (*facing page*), October 1978.
This was the view of his kingdom. Signalmen were practical railwaymen.

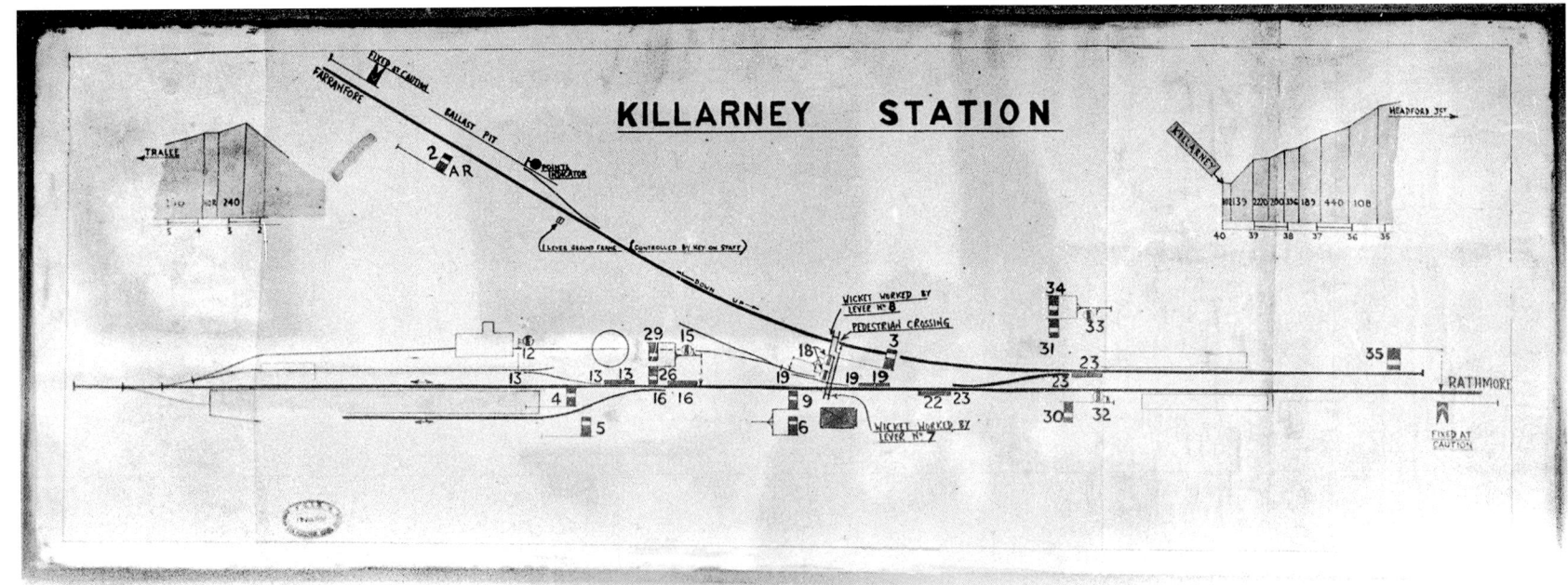

The signal box diagram with gradient diagrams included. On the Tralee line Signal 2 is an Automatic Repeater for Signal 3. That is to let the driver know whether or not he 'has the road clear' into the reversing siding.

B177 coming down the slope to reverse into the station, 2nd October 1974.

Looking towards the station, 4th April 1979. Killarney's down inner home signal, No.30, when lowered allows a train into Killarney Station. Disc 32, below, signals trains into the goods yard. On the right the highest arm, 34, signals out of the reversing siding to Tralee. The lower arm, 31, signals a train into the station. Below that, the square signal, No.33, with a yellow stripe, signals a train into the single line to go to the Ballast Siding. That siding was a sort distance out on the Tralee line; see diagram on page 10.

Dromore level crossing No.1 with the Keeper's cottage on the Farranfore – Firies road, 14th April 1979. This is about about 1,100 yards south of Farranfore Station.

Farranfore Station looking south to Killarney, April 1979. The overgrown land behind the down platform (on the right) is where the rails of the Valentia Branch used to be.

Farranfore Signal Box and Station looking north to Tralee from the up platform, 14th April 1979.

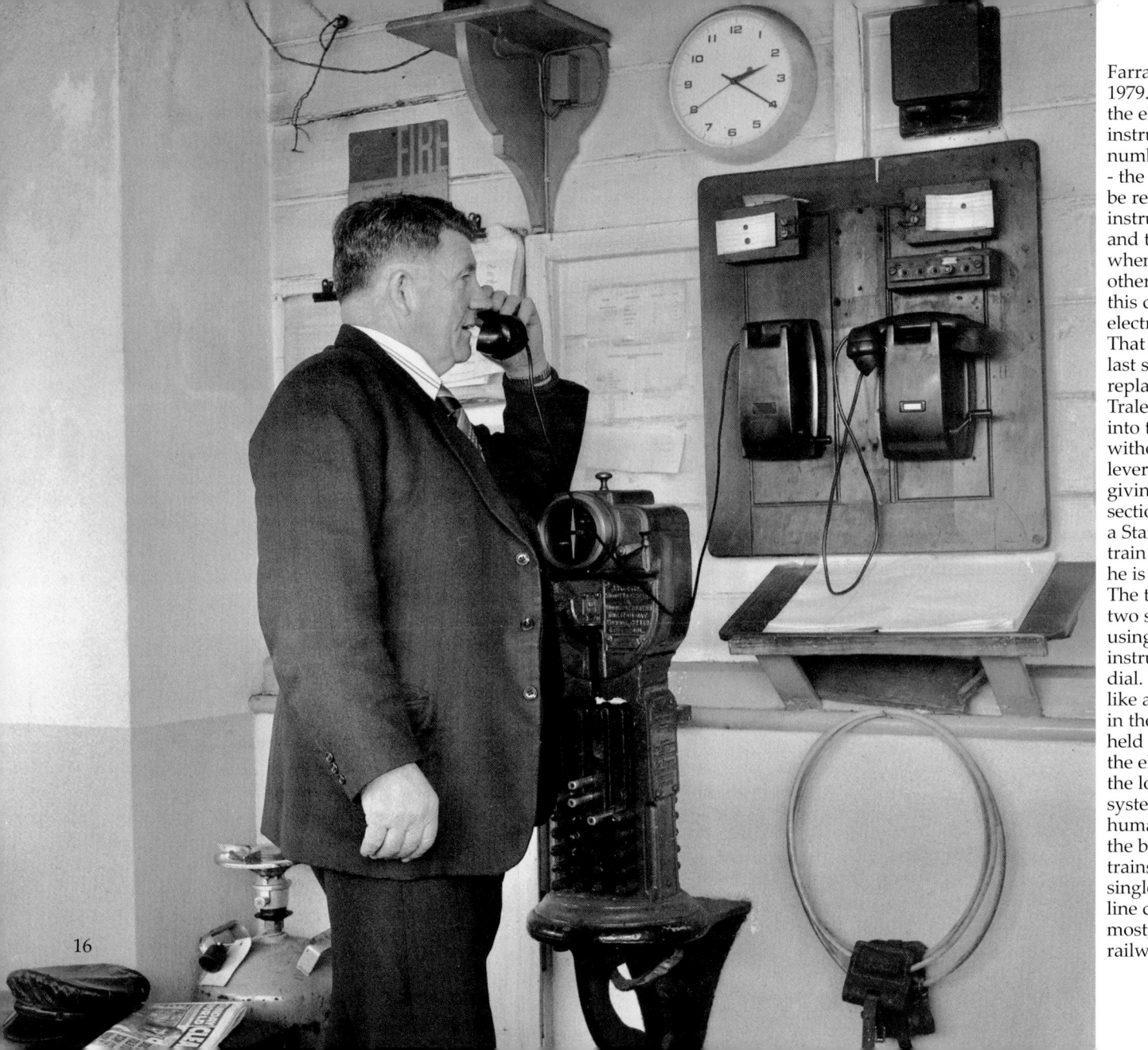

Farranfore signalman, April 1979. Behind the signalmen is the electric train staff instrument. It contains a number of small metal baton's - the Staffs. Only one staff can be removed from the instrument at any one time and that can be done only when the signalman at the other end of the section – in this case Killarney or Tralee – electrically releases one staff. That can only be done if the last staff used has been replaced in the Killarney or Tralee instrument or put back into the Farranfore instrument without being used. The signal lever operating the signal giving access to the next section cannot be pulled until a Staff has been removed. The train driver will not leave until he is in possession of the staff. The transactions between the two signalmen are carried out using the little lever on the instrument beside the large dial. Tapping the lever down like a morse tapper rings a bell in the next signal box, if it is held down for a few seconds the electric current will release the lock on the Staff. The system is as fool proof as is humanly possible and forms the best guarantee that two trains will not meet on the single line. This type of single line control was in use on most, it not all, Irish single line railways.

Farranfore looking to Tralee. A vintage latticework signal post on the left. The passing loop is signalled for working up trains over the down loop, if necessary. Looking carefully one can see where the Valentia Branch track split from the main line: that 2-arm signal would have been a 3-arm then.

Farranfore looking north from the south end, April 1979. Evidence of traffic that was there once in the big goods shed and the small jungle where once lay the tracks and the sidings for the branch line to Killorglin, Caherciveen and Valentia Harbour.

Farranfore looking south from the up platform. Is the signalman walking to meet his wife or an intending passenger wanting to buy a ticket?

A train at last! Coming in from Killarney, 14th April 1979.

And that was followed, after due process, by this respectable length of freight train.

Farranfore Station buildings on the upside. They have the look of a fortress. That seemed to be the case at a lot of stations.

The Laune bridge at Killorglin, 14th March 1979. The line from Farranfore to Killorglin was built by the Great Southern & Western Railway.(GS&WR) It was 12 ½ miles and opened to the public on 15th January 1885. This bridge over the River Laune consisted of a masonry arch on each bank of the river, those three Kerry limestone piers and the three 'bow and string' steel spans each 95ft. long. The route was extended to Valentia Harbour, 27 miles westwards on 12th September 1893. The line was closed on 30th January 1960.

Glenbeigh Station Signal Box, 13th April 1979, on the Cahirciveen/Valentia Harbour line from Farranfore and Killorglin. The gradients were mostly 1 in 50 for miles on this line. The GS&WR 0-6-0 tender engines must have been good steam raisers with expert firemen.

Kells Signal Box, 4th April 1980, on the steep climb from sea level up to the cliff edge summit section, through two short tunnels, high above Dingle Bay with breathtaking views across to the Slieve Mish mountains. John Power, who fired on the Dingle line and on this line, said that sometimes he and his driver saw the Dingle train, tiny in the distance, across the water, steaming down off the mountains and along the coast. John remembered both routes with great affection.

This is the down distant signal for Castleisland Station, 24th March 1975. The curious signal post was manufactured by Courtney & Stevens. The design was used on the Cork, Bandon & South Coast Railway and the Macroom Branch The branch line was 4½ miles long, leaving the Killarney – Tralee main line at Gortatlea Junction 7 miles south of Tralee. The line was opened on 30th August 1875 and purchased by the GS&WR in 1880. The GS&WR built a special train to work the line at their Inchicore Works. This was a coach body permanently attached as one vehicle to a tank engine, making what would otherwise have been an 0-6-0T into an 0-6-4 tank.

Castleisland Goods Station, looking into the town, 24th March 1975. The branch line was closed in February 1947. A goods train service was reinstated in 1957 and was permanently closed in 1977. The 0-6-4 combined coach and tank engine had long before been separated and the 0-6-0 tank engine was placed on a plinth at Mallow Station.

Tralee Station east end looking west, August 1976. The carriages in sight are at the bay platform built in April 1897. The middle signal routes to that track. The left hand signal routes to an older platform out of sight behind the bushes. That track led to an 'island' platform. The standard gauge track terminated alongside the north face and on the south face the Tralee & Dingle railway terminated with a 'run-round' loop for the T&D engine. The right hand signal routed passenger trains into the roofed over station. The track on the right of that was a goods line.

Looking across from the cattle pens to the Bay platform with coaches stabled. The two armed signal route top arm along Goods Loop, the bottom arm into the station. The Edward Street Signal Box visible between the tanker wagon and the box vans.

A very clean and decent Tralee Station, August 1976.

The handsome offices of C.I.E at Tralee. Beautiful stonework, similar to Killarney Station.

Driver John Power looks back for instructions at Tralee, 4th August 1976. He had invited me to come, with him to Listowel. I went with him next day. That long wall on the left is the remains of the walkway between the GS&WR and the Limerick & North Kerry Railway's terminus, opened in 1880. In 1901 the owners of the North Kerry line – the Waterford, Limerick & Western – amalgamated with the GS&WR making the North Kerry line terminus redundant. However, trains to and from Fenit continued to use it until 1913.

The 5th August 1976, riding in the cab of 017 with John Power on the 10.30 to Listowel goods. We had a 15 wagon train. We were running on the central track in the picture, approaching Edward St. Signal Box starting signals which were also Rock St. Signal Box home signals. – dual controlled. The signal is lowered – just about – for Listowel, the other signal would route to the Fenit line.

Rock Street Crossing 4th August 1976. Passenger trains Limerick – Tralee were abolished on 1st January 1963 but the Rock Street signalman was still needed while sugar beet trains ran to Abbeydorney and Fenit and ordinary freight trains to Listowel until 1978 when the line was closed. He had to hand out the Electric Train Staff (ETS) to Fenit and Listowel line trains.

The distant signal for Fenit Station, the wildest signal in the wild hills beside the Atlantic Ocean. Once there was a station further on, with a platform and a Station House and an office and an embankment out into deeper waters to allow the ships to come in to load and unload.

The down home signal for Fenit Station, a Saxby & Farmer product from before the First World War still standing proud, 22nd July 1977.

Fenit wagon load gauge. It made me think of a gallows for this remote, sad, ghost of a railway under the Slieve Mish Mountains.

Approaching the barriers at Ardfert going towards Listowel, 5th August 1976.

Abbeydorney Station on 12th May 1979. These stations were like a pleasant dream. Lost in a great wide landscape bounded by faraway distant misty blue mountains. I felt very lucky to have been invited. I think John Power knew well he'd been giving me a treat inviting me to come.

The signal box was open at Listowel and 017 and its train was welcomed by the home signal being cleared to 'All Right' to let us in and shunting ensued. Wagons of beer, cement, fertiliser and empty wagons required for loading were put off in different sidings and along the unloading platform. Then the train was re-marshalled for the return journey.

On arrival Driver Power and member of the station staff have a conference on what shunting has to be done

All the rail surfaces were bright steel from daily use. The up starting signal in the distance is very slightly cocked 'Off' for 017 to go out of the loop and back on the down line. Just a shunting movement.

Three vans rolling loose from the shove from 017 into the siding. I went to the signal box for a look around. The signalman was outside taking the numbers and details of all the incoming wagons and those going out. There were 22 levers working in the frame. When he came back we had a conversation. He feared being made redundant, and having to drive 42 miles each way to Tralee for his next job – if CIÉ could find him one. He worried that if he left the service he'd lose his pension.

A relic of steam days. Massively built to last 100 years and more.

With the shunting complete and the re-marshalled train standing at the platform, No 017 'ran-round' the train and came back onto the wagons facing for Tralee. John Power, his guard and myself went out of the station to the nearest pub to eat our sandwiches and drink a pint of Guinness.

Abbeyfeale Station, 22nd July 1977, looking as if a train might appear at any moment. There's a great feeling of loss and sadness looking at these ghost stations whether in Ireland or England.

Abbeyfeale looking towards the summit at Barnach Station, the latter set deep in that dark rock cutting.

Abbeyfeale looking towards Listowel. That 5 mph speed restriction notice, appearing to be so new, gives the impression the railway is still working here.